Happy Stats

Richard Evans

Published by Bremerhaven Press

Print ISBN 978-0-9963996-1-6

Bremerhaven Press
New York • Kiel • Amsterdam

About this book

I collected these guidelines over the years to help researchers save time and money. These are guidelines for working efficiently with me by avoiding miscommunication and trivial but time-consuming mistakes. Every statistician has different consulting styles, so use these guidelines to start a conversation about study design and data organization.

1.

COMMUNICATE DIRECTLY WITH THE STATISTICIAN.

Principal investigators, not minions, should communicate with the statistician. Minions can't make critical decisions about study design and analysis, so they have to relay questions to the PI. The back-and-fourth communication between the statistician and the PI using the minion as an imperfect courier adds days to a project. Fifteen year ago I kicked a minion out of my office insisting that she return with her PI. She did, and her PI and I became longtime collaborators, workout partners and close friends. I married the minion.

2.

DESCRIBE THE SCIENCE SUCCINCTLY.

Keep scientific explanations brief. However, the statistician might like to tour the lab or clinic and be included as part of the team. If practical, invite them. A few statisticians have become knowledgeable in science or medicine. If you can find one of those rare cases, nurture the relationship.

3.

GRAPH THE DISTRIBUTION OF EVERY VARIABLE.

Inspecting distributions helps identify wrong or confusing data values that can be intercepted before they are sent out for analysis. I was once asked to compare bandaging types over two groups, but all the dogs in the study had received exactly the same kind of bandage. I thought there must be a mistake in the data. It took several emails and a day to resolve. Had the investigator graphed the distribution of the bandage variable, he would have seen that it had only one value and dropped it from the data set and analysis.

4.

PUT VARIABLE NAMES IN THE FIRST ROW ONLY.

Statistical software reads the first row of a spreadsheet as variable names. The second row and beyond are read as data. If the variable label *patient name* is in the second row, then "patient name" becomes the name of a patient, which gums up the analysis.

5.

REMOVE CALCULATIONS FROM THE DATA.

Many times have I accidently included and analyzed column averages as a case because the researcher calculated them at the bottom of each column and then left them in the data set. Calculations, notes—anything left in the spreadsheet may be read as data.

6.

STATISTICAL SOFTWARE IS CASE SENSITIVE.

Yes, yes, and YES are considered three different outcomes. Beagle and beagle are different breeds. Annie Evans and annie Evans are different subjects. It's shocking how long it takes to hunt down those little problems.

7.

EMPTY CELLS ARE MISSING VALUES.

An n/a, period or other symbol in a cell is interpreted as data, not a missing value. So, for the variable *weight*, using "n/a" means that some subjects have a value of *weight*, which is not missing, called "n/a." Character values preclude calculating summary statistics, like averages and standard deviations.

8.

PUT NOTES IN NOTE COLUMNS.

It makes sense to include case notes with the data. For example, a radiologist may score radiographs and make notes about image quality alongside some scores. That's OK as long as the notes are in a separate column, a column of notes. If the notes are in the same cells with scores, then they are treated as scores. So, "3, image poor" is the score, not "3."

9.

USE ONE FONT STYLE.

Statistical software ignores font, font color, and highlighting so any meaning they have in the spreadsheet gets lost in translation to the statistical software.

10.

SEND A COMPLETE DATA SET.

Your statistician may want to check the format of the data set, but generally, keep the data set until all the data are collected. Interim analyses require preplanning to get the P values right.

11.

COMBINE MULTIPLE DATASETS INTO ONE.

Multiple spreadsheets, and the process of combining them, increase the number of errors. For example, I had three spreadsheets from three readers scoring the same radiographs. Each data set had the same variables but with some different variable names (*Case, case, case #*) and some different patient names for the same patients (Candy, Smith; Candy Smith; and Smith, Candy). It's safer and faster when the PI combines spreadsheets and then graphs the variables' distributions. See guideline No. 3.

12.

CHOSE THE RIGHT STATISTICIAN.

Statisticians subspecialize into refined topics and narrow practices: teaching, mathematics, survey sampling, big data, clinical trials, diagnostic tests, and so on. Would you like a dermatologist repairing your spine? Have a neurosurgeon fix your spine and have a biostatistician analyze your medical data. Experience counts too.

13.

USE CONSISTENT LANGUAGE.

The statistician will not know slang or acronyms, so use terms consistently. Puzzling out the difference between "extra cap" in an email but "TightRope" on the spreadsheet will add a day to the statistical analysis. The same goes for initialisms. It takes another day to find out that "MFS" in an email is the same variable as "Frankel" on the spreadsheet.

SUMMARY RULES

These rules distill to two: keep communications open and fast, and keep the spreadsheet clean and simple.

ingramcontent.com/pod-product-compliance
ning Source LLC
ersburg PA
060500200326
CB00017B/4870